GIRLS GUIDE: HOW TO COPE WITH SOCIAL MEDIA ADDICTION©

Angela D. Coleman, MBA, CNM

SISTERHOOD AGENDA®

sisterhoodagenda.com

First published in 2023
By Sisterhood Agenda Enterprises, LLC

Sisterhood Agenda Enterprises, LLC
Girls Guide: How to Cope with Social Media Addiction

administrator@sisterhoodagenda.com

ISBN-13: 978-1-952528-02-6

Printed in the United States of America.

GIRLS GUIDE:
HOW TO COPE WITH SOCIAL MEDIA ADDICTION©

Introduction

Maybe it's the mom in me. I want all of our girls to be healthy: physically, mentally, and spiritually. Every time I read a study about how social media is having negative effects on the lives of young people, girls especially, I worry.

At the same time, I don't want to be all doom and gloom about it. Social media is part of technology and tech just is. It's our reality. Technology is an important part of our lives, our worlds, and more often than not, our day-to-day routines depend on it.

But it can turn on you. When technology starts to consume you instead of you consuming it, it becomes an addiction. And like any other addiction, it can ruin your life and spoil your future. It becomes the thing you just can't quit.

There's no doubt that many of our girls are in crisis. Even before the pandemic, which did not help anyone's state of mind, girls were increasingly reporting being anxious, afraid, emotionally fragile, easily manipulated, or depressed. And although the link between technology and emotional well-being may not always be clear, studies have shown that the more hours that young people spend on social media, the more likely they are to experience disordered mental health.

The fun, convenience, and access that technology provides has made our lives so much better, but it can come with a cost that's often overlooked. Things like mental stability, emotional balance, and critical thought can suffer when the online world becomes a higher priority than the real one.
How many hours are too many? When does a habit become overconsumption, the precursor to addiction?

How can we consistently act in our own best interests? Being aware of what's behind your screen is the first step you can take to ensure that you're using technology wisely. You'd probably be surprised to learn the extent to which technology companies manipulate what you are seeing with the intent to direct your consumer behavior. They use sophisticated algorithms and gather your personal data while monitoring your online behavior. Often without knowing it, you give up your privacy, and the privacy of your friends and family, when you use many of these platforms.

If this sounds scary, it should. Fear can be a good motivator. And I want to motivate you to take control of your online life and not let social media play an unhealthy role in your life in the real world.

Sometimes it seems like technology is taking hold of our collective souls. Privacy, cell phone etiquette...these things are often nonexistent. We rarely go anywhere without our phones, and we use them all the time.

The excessive use of social media is not just a problem among young people. Most adults find the validation feedback loop of likes and attention to be irresistible too.

Our social media activity should not be what defines us. How we show up in real life is what matters. How we choose to engage, what we accept and ignore, makes a difference. Do we really need to access technology 24/7? Whether at a concert, while getting our nails done, in the car, at the dinner table, or even in bed – why is going without our phones for even 5 minutes becoming so hard to do?

If addiction to technology is the norm, what has our

society become? Technology itself is not the enemy; how we use it is what matters. Nothing is perfect and I believe that there is good in nearly all things. Social media, like all technology, provides us with incredibly useful tools. But we need to be smart with our social media use and keep it in check so it doesn't hinder us from flourishing IRL. Knowledge, having effective coping strategies, and moderation are the keys here.

Girls Guide: How to Cope with Social Media Addiction examines the impact that social media has on girls' development and on adults too. I will share psychological terms and social phenomena so that you can clearly understand how your addiction to social media is not an accident but a premeditated, well-orchestrated, and well-funded multifaceted strategy designed to gain money, information, power, and control for the corporations behind the content you consume online.

Perhaps most importantly, this guide should help you develop a more balanced approach to living your best life, one minute, one hour, one day at a time, with or without social media.

In the Spirit of Sisterhood,

Angela D. Coleman

OVERCONSUMPTION

"Who looks outside, dreams.
Who looks inside, awakes."
Carl Jung

It probably started innocently enough. Maybe you registered on a site to connect with friends and old classmates. Maybe you were seeking news on current events. Or showing off a unique talent or skill. Maybe you were just looking for something fun and interesting.

Is social media fun, cool, interesting, inspirational, and educational? Is it manipulative, deceptive, mind-altering, and soul-sucking? It can be neither or all of the above. Of course, it can be anything because it is what you make of it and allow it to be. It can be a big lens or a small one, and each generation may have a different perspective.

Whether you see your social media use as a vortex of insecurity or a safe haven for light-hearted distraction, somehow, over time, it might have become a time-consuming habit that stopped serving your highest good. Instead of twice each day, maybe you started checking in every hour or so until it became something that you felt compelled to do, no matter what, all the time, for at least three to five hours each day. What was once entertaining and informative may have easily become an overwhelming need that took priority over the other things that you know you needed to do in your life. Can you imagine living without it?

The Nature of Addiction
Social media enlightens, entertains, connects, and creates community in an online environment. In and of itself, it is neither "bad" nor "good."

Social media can be a powerful teaching tool and one that creates a vibrant and healthy form of self-expression and unity. But like almost everything that can be good for you, doing too much of it can become a problem. Overconsumption of social media can represent an addiction, even though it has not yet been identified as a clinical disorder.

Most people who do not understand social media addiction do not understand the true nature of addiction. Do you think it's only drugs or alcohol you can be addicted to? That's a common misconception. It is a fact that you can be addicted to pretty much anything, as long as you feel like you are being rewarded in some way by the addictive behavior, even while it disrupts your life.

Photo credit: Becca Mchaffie-Fzde, Unsplash

Social media addiction is classified as a behavioral addiction, like shopping, gambling, and sex addictions. You can become addicted to a substance, a person, a thing, or an activity. Anything enjoyable that is best consumed in moderation can become an addiction.

Denial is usually very strong with all kinds of addictions. Sometimes, the addicted person doesn't realize what's wrong, only knows that her life isn't super-great right now.

Addiction is chronic, meaning the addictive behavior happens often enough, usually every day, to cause disruptions in your important relationships and daily routine. Your daily activities are centered around your need to feed the addiction. In short, addiction causes problems in most or all areas of your life.

Social media can cause a spike in dopamine, a feel-good hormone, which explains why it can be addictive. Like a mouse receiving a food reward when it presses a button, you often get small rewards when you find something interesting and stimulating on a social media site, get a like, or receive an affirming comment.

Do you have an addictive personality? An addictive personality makes you more susceptible to addictive tendencies because you crave rewards over and over and feel a need to repeat experiences that feel good in the moment. What you get is never enough and you feel like you always want *more*.

Social media addiction shares characteristics with other types of addictions. According to Goodrx.com:

"Substance use disorders and social media addiction are similar in that both can involve tolerance and withdrawal. Like people with substance use disorders, those with social media addiction can develop a tolerance, where they crave more social media time to feel as good. They can also experience withdrawal symptoms when they're not on social media, like irritability and anxiety."

Social media addiction, in particular, is correlated with anxiety and depression (more on this topic in the next chapter), lower self-esteem, decreased social skills, confusion regarding personal identity, and conflicts related to your divine purpose.

Warning Signs

Do you feel isolated and invisible? Disconnected or unsupported by your community? Maybe you think that you're fine. Home alone in your bedroom, your couch, in a little café, you are in your own small bubble of comfort, convenience, and affirmation. "I am just getting information and learning," you tell yourself. "Everything's fine." That's what you may say. That's what you may believe. It may not be true. You could be living in a bubble.

Photo credit: Kenneth Surillo, Pexels

Do you hide your social media use from family and friends? According to a Pew Study, 70% of girls hide their online behavior from their parents. Anyone who spends so much time with very little to show for it may be stuck and lost in some way or in many ways, even if she doesn't see it.

By recognizing the signs of social media addiction, hopefully you open your eyes to the dysfunction that may plague a large part of your life. Warning signs of a social media addiction can include all or any of the following:

- Denial of addiction
- FOMO, Fear of Missing Out
- Lack of control and inability to stop using social media
- Lost productivity and time
- Dependence on social media for information and feelings
- Preoccupation with social media, even when not in use
- Changes in appetite and sleeping
- Decreased socialization, ignoring important relationships
- Disregard for the harm caused to yourself and others
- Obsessive and compulsive checking of social media
- Loss of interest in real-world activities
- Insomnia and altered sleep hours due to social media use
- Negative change in physical appearance: lack of hygiene
  and grooming, weight loss/gain
- Changes in personality
- Increased irritability
- Delusional and irrational thinking
- Paranoia
- Decreased attention span

- Globalized anxiety and depression; the world feels dangerous, unsafe, and unfair

In summary, social media use can be very difficult to limit or stop. Most girls report high levels of confidence in their ability to manage their settings, even when they are addicted and they actually lack this control. Other signs of social media addiction:

- You do it in the morning as soon as you wake up, or it wakes you up.
- You do it at night before you go to sleep.
- Whenever you have a free moment, you do it.
- You tell yourself that you can stop at any time, but you don't ever stop.
- You'd rather do it than anything else.
- You find more meaning in the likes and online conversations than in real-life relationships and conversations.
- You stop only to get food or to go to the bathroom and even then, you often take it with you.
- You spend most of your days indoors and alone, choosing to order in instead of going out or making your own food.
- You eat your meals with your phone.
- You make up stories about what you do with your time instead of telling the truth.
- You use the "everybody's doing it" justification.
- You are very defensive when questioned about it.
- You are convinced that "they just don't understand."
- You vastly underestimate how much time you spend on it.

Photo credit: Shutterstock

Studies show: the greater your social media use, the less motivation you have to improve your life and the less likely you are to reach out to friends and family in person. If it lasts for a long time, social media addiction can turn your life into a vapid, numbed existence that lacks psychic and emotional depth. This may sound dramatic, but it's not just a worried mom's exaggeration – it's the truth.

Your Truth

Conquering social media addiction means embracing your truth. And some of us know all too well how much the truth can hurt. With courage, you can choose to move forward anyway.

Speaking of truth...social media is not always honest. And some of us have gotten used to the dishonesty. However, I encourage you to honor yourself by giving yourself the complete, 100% truth. The truth about the number of hours that you lose each day, the truth about what you are reading and watching, the truth about the reward, and the truth about your "why."

Answering your whys is important to help you grow out of social media addiction. Why do you feel the need to share and engage with strangers? Why can't you put down your cell phone? Why do you seem to enjoy trafficking in the trauma and tragedies of others? Why do you care what people who don't know or care about you think? WHY?!

Where does the desire to live publicly and share detailed aspects of our lives with strangers come from? Why and how do some of us so readily disclose without trust? What makes us want to record our story, even when we know what we post online can be used against us?

Are you bored? Lonely? Most of us crave community and we need an outlet to express our unique talents and divine gifts. That is the "social" part of social media. The feeling of belonging to a select group feeds our need to feel special. This is a natural need and is easy to understand. Social media can be all about our feelings.

Emotions play a major role in motivating our behavior. We do what we like. With its headlines and promotion of feelings over facts, social media use may start off as a very healthy way to get information, learn more about other people, and express your point of view. Yet today, it could feel like you are stuck in an endless stream of content, more to look at, more to see, more views and likes to get...you turn to it often, every day.

As much as addiction makes people's lives miserable, we also have to be honest about what you get from participating. The "feel-goods" that you get from engaging in social media are real. And this is why we do it, why it's everywhere, all the time.

The Business Side
Social media is a business and going viral can be very financially rewarding. If you desire to do something grand in this life, especially in business, social media can help you achieve your dreams. Your dreams, whatever they are, should be supported, but not at any cost.

Let's encourage dreams, not addiction. What if social media causes fixation and addiction by design? Social media companies work really hard to keep you on the site, keep you engaged, and keep you in your feelings for as long as possible. And the negative feelings are just as relevant to them as positive ones because it is an AI-generated, algorithm-based monetization scheme in full effect.

Corporate greed rules the day and social media companies generally do not prioritize your mental health. They are only interested in the money generated from you being on the site; the more you join, watch, like, subscribe, comment, and click, the more money they get. And the more data they gather, the more effective they become at doing what they do.

In many ways, what the companies are doing interferes with your capacity to share, engage in your truth, and maintain positive mental health. However, it is something that can be overcome.

Reach Out
Social media doesn't have to hurt you, and you do not have to be addicted to it. When you go overboard with it, it can cause you to feel bad, block your blessings, and delay your destiny.

You don't have to suffer. Addiction is treatable, but you've got to be willing to do the work. Reaching out for help may be uncomfortable at first but your bravery will yield rewards that benefit all aspects of your life. If you are reading this book, your transformative journey has already begun.

Photo credit: Matheus Ferrero, Unsplash

Some experts believe that your environment can cause and trigger addiction. Others point to genetics as playing an important role; for example, if your mother or father has addictive tendencies, it is more likely that you will have them too. However, just as girlhood trauma does not have to define your life (see *Life After ACEs: Overcoming Girlhood Trauma*, another of my books in this series), addiction is not destiny.

This is worth repeating: addiction is not destiny. Your brain and life experiences are part of your addiction, and although they may have been encouraging your addiction, now they can be redirected to encourage your recovery and progress instead.

Recovery

"Recovery" is a term used to describe the process of enhancing holistic health and wellness through positive behavioral change, including the reduction and eventual elimination of overconsumption activities. SAMHSA, the Substance Abuse and Mental Health Services Administration, defines recovery as: "A process of change through which individuals improve their health and wellness, live a self-directed life, and strive to reach their full potential."

Photo credit: Cottonbro Studios, Pexels

Psychotherapy of all kinds, especially behavioral therapy, can help a lot with recovery from addiction. You have the power to heal yourself; it's just a matter of looking inside yourself to find it.

The quote that began this chapter acknowledges the dream-like state that looking outside of yourself creates. In this way, engaging in social media is like sleeping and daydreaming, falling down the rabbit hole like Alice in

Wonderland. Social media addiction can be like being trapped in a techno trance.

Just like we can sleep, dream, and exist in a trance, we can also choose to wake up and live.

Like Spike Lee said in the movie *Do the Right Thing*, "WAKE UP!" Are you willing to take the red pill and see your reality like Neo in *The Matrix* or continue to lie to yourself each day that this is all life has in store for you? Or do you want to take back control of your attention and your life?

What You Can Do:

- *Be honest about your social media use. Give yourself 1 point for each warning sign of social media addiction listed in this chapter that is true for you. If your total is more than 10 you are likely addicted to social media.*

- *Monitor your social media usage on all the sites you visit regularly. Add up the time throughout a 24-hour period to get an accurate picture of how many hours you are spending on social media.*

- *How do you feel when you are on your favorite site? Think about how you are rewarded by your engagement and what makes you stay on it for a long time.*

ANXIETY AND DEPRESSION

"There are wounds that never show on the body that are deeper and more hurtful than anything that bleeds."
Laurell K. Hamilton

Most girls want and need to be seen, heard, and accepted. These needs and the desire for connection and approval can drive social media addiction. At first it may seem like a safe escape and even feel like you are in control, but it can quickly become apparent that too much online scrolling can be addictive and unhealthy.

There are signs if we pay attention. But first...ask yourself where the desire to escape comes from? What are you escaping from? Your life? Your success? Work? Progress? Family? Friends? As previously stated, addressing the root cause of your behavior means identifying your why.

That fuzzy feeling of comfort, safety, and familiarity you get with your social media family feels warm and easy because it is made to feel that way. I want to (gently!) challenge the idea that it's safe. Rather than thinking of it as your comfort zone, let's look at it more like sleepwalking. When you are sleepwalking, you are not fully aware of your surroundings; you could even step off a cliff (people reading their cell phones have actually done this). That is not safe.

For some people, there's no sugar-coating it: social media addiction is downright self-destructive. Sound extreme? It's not at all. Psychologists define self-destructive behavior as a pattern of actions that could hinder a person's physical and emotional well-being or their ability to achieve important goals. Self-destructive behaviors inhibit personal growth and lead to physical, mental, and emotional issues. Self-destructive behaviors can be very obvious or really subtle.

Examples include:

- Drinking too much
- Drunk driving
- Putting yourself in risky or dangerous situations
- Ignoring bodily needs for movement, nutrition, hydration, etc.
- Pulling continuous all-nighters or otherwise denying yourself adequate sleep
- Mindless or endless social media scrolling
- Watching too much TV
- Engaging in risky sexual behavior
- Cycling through toxic relationships
- Cheating when you're in a happy relationship
- Hurting people you love
- Denying your own needs or wants
- Procrastinating
- Taking on more than you can possibly handle
- Excessive spending or spending money you don't have
- Endless ruminating or dwelling on past mistakes
- Negative self-talk
- Impulsive decision-making
- Isolating yourself from your community and loved ones
- Putting yourself in situations that are traumatizing or re-traumatizing
- Ignoring mental health symptoms like anxiety or stress
- Self-harm
- Suicide

Source: MindBodyGreen

As explained above, the damage done by social media is not just done to your physical self, at least not right away.

The real harm is psychological and mental. The effects of social media addiction are often unseen and therefore difficult to measure and, for many people, difficult to understand. However, know this: the wounds created by social media addiction are very real, including increased anxiety, depression, isolation, and loneliness. And there are often scars, long-lasting impacts long after the actual social media use is over.

Social media has the capacity to lift you up but it can also reduce your spiritual awareness. Because your brain is being teased and manipulated, your common sense and intuition may decrease. Your intelligence and critical thinking skills – right out the window! What happened?

Photo credit: Christina Wocintechchat, Unsplash

Compromises

Social media companies don't want you to use your intelligence and critical thinking skills. This "dumbing down" is on purpose. It makes you more vulnerable and ready for the sales pitch. What sales pitch? Look closely and wait for it... it's always there.

Social media and commerce go hand-in-hand. Social commerce is a thing. That is the compromise. It's a big one. With social commerce always around, our capacity to be free-thinking and form original thoughts is repressed. As a result, we become more susceptible to suggestion, more naïve about who is an "expert," and more open, in general, to everything, especially influencers. Influencers command a $16 billion-plus marketing economy.

When we lose our critical judgment, we lose freedom, usually without realizing it: freedom of thought, expression, and action. The group mentality becomes more important than your own and there is intense peer pressure to be agreeable with the masses, to go with the flow. You go around repeating what they "know" and sharing what someone said on social media instead of investigating the issue yourself. It's a pretty lazy way to live, and becoming more and more common. Do you know someone like this? Is it you?

People and events become tantalizing headlines, reduced to clickbait. Feelings (the more intense, the better) become more important than actual content or knowledge. Being "right" has nothing to do with stating and understanding facts. People reading only headlines and comments tend to believe that they are more informed than they actually are. We can lose information and the will to process it.

Trauma

It is common knowledge that social media can reinforce trauma; it can traumatize, trigger, and re-traumatize. Trauma dumping, bringing up traumatic events in casual conversation, is also common on social media. This is a big risk for women with ACEs (adverse childhood experiences).

Is your social media addiction related to childhood trauma? In my book *Life After ACEs: Overcoming Girlhood Trauma*, I write about how everyone responds to trauma differently. And those responses can be unpredictable, appearing even long after the traumatic event has occurred. Constantly engaging in the traumas and tragedies of others, even from a technological distance, has many effects on us, including:

- It can trigger memories of your own traumatic experiences.
- It makes you feel better about yourself and your life, in general, because you see someone having a more difficult time than you.
- It is interesting, even entertaining to you, on some level.
- It feeds into voyeuristic tendencies to spy on other people, learn and see something that others do not usually know.
- It can cause a response (positive or negative) that you can immediately feel and respond to in real time.
- It can make you addicted to drama in your own life.
- It can blur the lines between fiction and reality and cause you to mistake other people's real-life experiences for a fictional movie.
- It can reduce empathy and appropriate emotional responses.

It is no wonder that many studies show people who engage in frequent social media use also report high levels of anxiety and depression. This very important finding makes sense in light of what we know about how these technologies are designed to draw us in.

The psychological effects are especially troubling when it comes to impressionable young people, whose brains are still developing and whose minds are often struggling with

Photo credit: Medium Photoclub, Pexels

personal identity and issues related to culture, race, gender, and sex.

When too much information is received too quickly, the brain becomes overloaded and overstimulated, which increases impulsivity and decreases the ability to focus.

What you may not see in the statistics about social media is the unrelenting onslaught of pain, criticism, and difficulties that people are experiencing, especially vulnerable young girls who are more likely to suffer from cyberbullying mean girls.

Symptoms

Most girls have a hard enough time growing up without social media. With their lives and the lives of others in the spotlight, there is so much pressure for idealized public presentation. Understandably, the pressure to present a perfect image creates anxiety.

Symptoms of general anxiety include:

- Accelerated heart rate, high blood pressure
- Excessive sweating
- Nausea, upset stomach
- Tight muscles, knots
- Fatigue, general weariness
- Overthinking
- Negative self-talk
- Fear of outcomes, worst-case scenarios
- Inability to let go of worries and fears

Symptoms of general depression include feelings of:

- Hopelessness
- Helplessness
- Sadness
- Isolation
- Restlessness
- Indecisiveness
- Fatigue
- Emptiness
- Insomnia and/or oversleeping
- Self-harm, suicide
- Physical aches and pains
- Digestive issues

Photo credit: Cottonbro Studio, Pexels

Are you feeling any of these symptoms on a regular basis now? What other feelings are happening right now? List them here:

____

____

____

____

____

Source: Girls Guide: How to Relax

Coping Mechanisms

The pandemic increased our use of technology to adapt to office closures, global isolation, fears, uncertainties, and changes that we were all experiencing. Public health mandates advised us to stay indoors and being online all the time was normalized. Many of us were doing the majority of our living and working online. In this way, technology provided a type of sanctuary during a scary time. Safe. Comfortable. Easy.

Photo credit: Andrew Neel, Unsplash

As anxiety and depression increases, so do coping mechanisms. No one blames you for wanting, even needing, a safe retreat sometimes. But...let's make sure your retreat is actually safe. And healthy, one that doesn't hurt you or put you in harm's way. Healthy coping mechanisms help make your life better. On the other hand, unhealthy coping behaviors may feel good in the moment, yet actually end up making things worse.

It is no wonder that rates of anxiety, depression, substance abuse and other addictions shot up at this time as we individually and collectively struggled to find ways to handle quarantine, shut-downs, health scares, and mass casualties. Addictive behaviors are an almost expected outcome when there is rapid change, confusion, uncertainty, pain, despair, and most of all, fear.

Fear is a big one. When you are afraid to be in the big, bad, scary world and venture out into the unknown, you stay inside the small one that you created for yourself. Here, you feel more powerful and in control, even when you are "doomsday scrolling."

Photo credit: Jermaine Ulinwa-3231773, Pexels

Research has shown that the link between depression and social media addiction is clear. Even though they are two separate and distinct behavioral disorders, they often co-exist. Additionally, depression can mask social media addiction and social media addiction can mask depression. It is possible to have one or the other, or both, with one being more dominant or both being equally affecting.

Social media addiction is so new that the medical, academic, and psychiatry fields are still struggling to catch up to the technology. Regulation, especially for children, also lags behind our current reality.

Photo credit: Bruce Mars, Unsplash

Go Deeper

When we go beyond our eyes and ears, we see the depth of our true reality. Everything, even you (especially you!), is so much more than it may appear.

Online, you may be part of that small community of believers on that social media platform regarding that particular issue at that particular time. It is not the entire world, although it may feel like it at the time. Going beyond the superficial, you may intellectually understand that this connection can be severed in an instant at any time with the click of a button.

Understanding that these "friends" and "likes" can change and turn on you, cancel you, make you feel bad and insignificant and even more alone than you felt before should make you question your dependence on them.

Social media can be unhealthy and confusing, and hazardous to your mental health and emotional well-being.

Is there soulfulness in social media? Some people believe so. But the behavior of cowardly keyboard warriors that bully those that they perceive as weak makes me question the platforms that give people online anonymity to say and do unthinkable things.

Soulless youth grow up to be soulless adults who, together, create a soulless nation. It's a bit frightening, isn't it? But we can stop giving our power away.

What You Can Do:

- *Be aware of false and inappropriate comparisons. Stop comparing yourself to others.*

- *Think of the role trauma plays in your social media addiction. Are you helping or hurting yourself?*

- *Recognize that the struggle is real. Are you suffering from anxiety or depression? If so, don't delay: talk to a mental health professional who can help right away.*

THE PSYCHOLOGICAL AND THE PHYSICAL

"Our own brain, our own heart is our temple; the philosophy is kindness."
Dalai Lama XIV

If our brains and hearts are our temples, we must take good care of them, right? We need protection from anything that may cause harm. And we can do this with kindness, kindness with ourselves and others, to keep us from feeling bad.

Photo credit: Christina Morillo, Pexels

What if your social media use made you feel bad? It would make sense for you to stop using it, wouldn't it? Who says social media makes teen girls feel bad? Teen girls do. Instagram data from March 2020 shows 32% of teen girls said that when they felt bad about their bodies, Instagram made them feel worse. An earlier report from 2019 said teens blame Instagram for increases in the rate of anxiety and depression.

Most teenage girls, especially those who have experienced trauma, share a desire to be noticed, which social media can help them achieve. However, it's also a time of life when many young people feel awkward or uncomfortable about their bodies, and through distorted lenses and filters, social media can make them feel worse. It's an unkind and unhealthy cycle.

Importantly, social media addiction can make us feel bad in ways that we may not yet quite understand.

Suicide Risk

The Atlantic reported surging rates of depression, anxiety, and self-injury for adolescent girls in the early 2010s, as social media platforms proliferated and expanded:

"Much more than for boys, adolescence typically heightens girls' self-consciousness about their changing body and amplifies insecurities about where they fit in their social network."

Suicide risk among teenage girls is rising fast and many experts point to social media as the cause.

Every three years, the Centers for Disease Control (CDC) conducts a survey. In 2021, the CDC saw an increase in mental health challenges among many people, "but it's girls in the U.S. that are engulfed in a wave of sadness, violence, and trauma." Nearly three in five reported feeling persistent sadness and hopelessness, 25% of girls reported having made a suicide plan and 14% reported having been forced to have sex. LGBQ+ teens, the study concluded, continue to face extremely high levels of violence and mental health challenges.

At the end of the day, we are all humans, but that might not be true on the computer. We do not know if bots, AI, or really unkind, sadistic trolls are on the other side of the screen, part of our "group," contributing to our social media community.

Are you struggling with suicidal thoughts, or do you know someone who is? Contact the 988 Suicide & Crisis Lifeline

Don't be ashamed to reach out for help.

Social Comparison

"When I see girls around my town that look really good on Instagram it makes me feel bad, like it makes me feel very insecure," said Emily Silva, a teenage girl interviewed in a news article last year.

Photo credit: Nick Arnot, Unsplash

Social comparison, the process of evaluating ourselves by comparing ourselves to others, is alive and in full effect.

It's *social* media, after all. Let's think about it a bit more. Most times, social comparison happens on an unconscious level, meaning we do not think about doing it, we just do it.

We compare grades, attractiveness, abilities, careers, status, pets, and anything you can think of. When consistently comparing yourself to other people on social media, you will always be at a disadvantage because what you see on social media is not real.

But...what if there are errors in your social comparison? Sociologists call it "erroneous social comparison," where you compare yourself to someone in a way that makes no sense and leaves you feeling deficient. Similarly, I challenge your concept of reality as it relates to social media. Do your eyes deceive you? Yes, they do. Social media deceives you and everyone else too. Your actual life will never be as great as a perfectly curated social media feed because no one's life is perfect.

Dr. David Greenfield, the founder of the Center for Internet and Technology Addiction in Connecticut, has said of this unhealthy way of thinking: "It's not a realistic representation of life. So, we're comparing ourselves to something that really isn't real and that is what creates a lot of the problems."

Your Shadow Side

Carl Jung, the psychoanalyst quoted in the beginning of the previous chapter, wrote about the shadow self, the part of ourselves that we repress and hide from the public because we rarely like to acknowledge it, even to ourselves. Everyone has a shadow self. It is not evil or bad; it just is. Your shadow self can be a source of inspiration and creativity when you become more self-aware, integrate what is hidden into your life, and use it to transform your life for personal growth.

Social media often encourages black-and-white thinking where you see things as either/or, with few nuances or exceptions to the rule. Reality is often much fuzzier and more circumstantial with shades of gray. When you ignore your innate complexities, you are hiding most of your true self, including but not limited to your shadow side.

Objectification

Objectification is when you take something organic and complex – like a person – and treat it like a "thing" or an object. This happens all the time to women and girls, both online and offline. But did you know that your social media use increases self-objectification?

This means, as you increase social media engagement, you begin to see and view yourself more like an object or "thing" than a real person.

According to a recent article in *Psychology of Women Quarterly*, as women engage in more strategic self-presentation on social media platforms, their reported levels of self-objectification also increases.

Okay. "Strategic self-presentation" is a fancy way of saying that you only post and share what you think shows the best of who you want to be. Many times, this is a persona, a tiny bit of who you actually are. The persona is not real; it's created using filters, photo-editing software, and other enhancements.

Photo credit: Anastasia Kolchina, Pexels

Groupthink
Rather than think independently for yourself, you may succumb to groupthink, the process of accepting what the majority of your social media group think automatically, often without question. Investopedia describes groupthink as: "a phenomenon that occurs when a group of individuals reaches a consensus without critical reasoning or evaluation of the consequences or alternatives."

Groupthink is based on a common desire not to upset the balance of a group of people. It's usually very unhealthy because it overlooks the needs of specific individuals.

Photo credit: Clarke Sanders, Unsplash

In essence, you are being told what you should think because it is what the rest of the group thinks. Groupthink stifles individuality and creativity and makes the individuals in the group feel less accountable than they really are. Groupthink is often based on feelings and the desire to belong rather than facts.

FOMO

Have you ever started off with an idea to respond to a comment and suddenly find yourself eating your dinner while still tapping on the phone or computer two hours later? In this way, social media can be a time suck, all because you have FOMO (Fear of Missing Out).

Social media doesn't eliminate the fear, just masks it, like a blanket over a wound. With FOMO, it can help to ask yourself, what am I so afraid of? What is the worst that can happen if I sign off?

Photo credit: Adrienn, Pexels

When you are on online, if you are not creating anything, focusing on your purpose, or busy developing yourself to fulfill your destiny, you have to wonder if you are literally wasting time. Often you may feel "busy" but this is part of what fuels the addiction. You may spend a lot more time online than you realize, time that you can never get back. Sometimes you just zone out.

Dissociative Disorder

Studies show that people can enter a dissociative state when using social media. As if you're sleepwalking, social media can keep you completely absorbed to the point where you are not able to pay attention to what is going on around you. Users report "spacing out" on social media platforms, along with the inability to remember what they just read. Sleepwalkers often do not remember their sleepwalking, either.

Photo credit: Angela Franklin, Unsplash

Dissociative disorder, like PTSD, is usually related to an unresolved trauma from childhood. When the mind is really stressed out, this is its way of coping. In this way, the person "checks out" because life circumstances have become too much.

The Mayo Clinic defines dissociative disorders as mental disorders that involve experiencing a disconnection and lack of continuity between thoughts, memories, surroundings, actions, and identity. There are many types and people with dissociative disorders escape reality in ways that are involuntary and unhealthy and cause problems with functioning in everyday life.

Symptoms of dissociative disorder include:

- Feeling disconnected from yourself and the world

- Forgetting about certain time periods, events, and personal information
- Feeling uncertain about who you are
- Having multiple distinct identities
- Feeling little or no physical pain

A medical professional can help with diagnosing any difficulties.

Photo credit: Rodnae Productions, Pexels

Narcissism

Do you dream big? Many high schoolers today may graduate with aspirations to become social media influencers. Influencers often wield great power, which can be used for good or not-so-good purposes. And it comes with conditions.

Photo credit: Paul Hanaoka, Unsplash

Maybe it seems easy. Or this type of job appeals to your independence, self-centeredness, or innate narcissistic tendencies. We all have our own motivation. But while striving for influencer status may appear easy, it takes a lot of effort and a lot of energy. Posting often, commenting, strategizing, promoting, responding. This type of work can be expensive and difficult to maintain in terms of mental and emotional energy – for many, it's just exhausting. Most influencers eventually suffer from burnout.

Also, this type of influence is usually unstable, fleeting, and subject to change, depending on the whims of the social media

platform corporation, your community, or anything else. Just as easy as it is to connect, they can disconnect just as easily, too. While being an influencer seems powerful, it rarely empowers the influencer.

When you dream, I want you to dream big, and in terms

of both virtual and real-world achievements – you deserve the best of all the world has to offer!

Psychological First Aid

Let's talk about psychological first aid, a useful therapeutic tool to promote healthy coping behaviors when you are triggered. Immediately available, psychological first aid is like emergency mental health care.

Psychological first aid is responsive to the most immediate needs that people have for care and support. Often used after natural disaster events, psychological first aid taps into the resilience of the human spirit and actively prevents post-traumatic stress disorder (PTSD).

Unfortunately, social media offers no psychological first aid. There is not even a warning label. Maybe these regulations are coming, especially for children. But for now, without support, you are left to figure it out on your own if you are triggered on a site. I suggest that you and your friends learn psychological first aid to learn how to support each other if and when you encounter something upsetting online.

And don't forget: immediate mental health counseling is available in the U.S. by calling 988, the national Suicide & Crisis Lifeline.

Bullying

Online trolls and bullying can be very aggressive and debilitating and has been documented to cause suicidal thoughts and actions. With what you now know about groupthink and suicide, this makes sense. But a little-known fact: social media addiction can also encourage you to participate in group bullying behavior in which you are putting others down. This happens:

In order to maintain your group identity
To allow you to project your own insecurities
To help you feel better about yourself

You might feel superior – for a minute. Bullying others can keep you from experiencing your biggest fears of rejection and being made fun of because you are the one rejecting and making fun of others. Some may argue that gossiping is also a form of bullying.

Photo credit: Nylos, Unsplash

Hurt people, they say, hurt people. Bullied people tend to bully others. Yet, you can break free and be more. I know that you can do better. By making smarter, more informed choices, you can resist the temptation to gossip and bully. It's not a good way to be, and karma has a way of balancing these things out! With self-awareness and more tools in your toolkit, you can protect yourself from being a victim too.

Cognitive Impairment

Research has shown that screen time should be avoided for kids under 2 years because it can cause negative cognitive outcomes in that age group. It changes their brains. The brains of middle school children (usually ages 11, 12, and 13) are very susceptible to outside influences that affect dopamine surges and self-perception during this critically important developmental stage in early adolescence. Dopamine surges means your brain starts wanting more, more, more!

It's not just the brains of very young children that are being impacted. Social media may be changing your brain, too. People's brains are not fully formed until at least the mid to late 20s. Even as an older adult, your brain can still be adversely affected by social media. The jury is still out when it comes to determining just how much social media affects the brains of people of all ages, in both the short term and the long term.

What You Can Do:

- *Note when and where you may be triggered on social media. Think about ways to prevent it from happening and ways to handle it if and when it occurs.*

- *Monitor your feelings when you are using social media. Do you dissociate? In what ways?*

- *What do you do when you encounter cyberbullying? Have a plan of action to disconnect, disengage, and get support if someone does that to you.*

DIGITAL DETOX

"I just don't like this compulsive, instantaneous, over-information, lack-of-privacy, weirdo aspect of the world."
Drew Barrymore

Living a great life online is great as long as you are also living a great life offline. Even so, every once in a while, everyone needs a break from technology.

Can you envision what a digital detox looks and feels like?

"Detox" is a term that means abstaining and refraining from toxic, unhealthy substances and the process of ridding your body of these substances in order to improve your health and well-being.

In this case, we are viewing social media over-engagement as the toxic, unhealthy substance. By refraining from engaging in social media, even for a short time, you can sever its hold on you and take your power back, the power that you gave to other people to have control over your life.

Did you give away the power to judge you and provide intimate details of your existence to complete strangers? Did you give big corporations and advertisers permission to spy on you and access all of your contacts? It's never too late to take it back. You can. You will if you want to recover from social media addiction.

Break Free
Breaking free means no longer allowing yourself to be a prisoner in that social media bubble that you created on those sites. Maybe you think you need it. But you don't. Breaking free means unplugging, disconnecting, and

enjoying the space that you create to do and be whatever you want. Breaking free is knowing that you are all that you really need. You had a life before your social media addiction and you can have an even better life now by taking a break, even temporarily.

Photo credit: Matheus Frade, Unsplash

Breaking free releases the tight grip that social media has on you. It means truly being your own person, using your own mind and thought processes, and being clear about media manipulations.

Above all, it's about positive mental health, well-being, and self-esteem. It means embracing a lifestyle that is very different from your current one and may be different from your peers and others you know. It means creating new space to improve the quality of your life. It's what we need to truly be healthy. It's not our fault that many of us are addicted to social media. It has been designed to get us hooked.

Slow Down

Dr. David Greenfield, founder of the Center for Internet and Technology Addiction and author of the book *Overcoming Internet Addiction*, called social media "digital fast food."

Instead of fast food or quick, instant highs and lows, let's take our time. Your health will thank you! Instead of rushing full speed ahead, slow it down. Take your time. Relax and make deliberate efforts to move more slowly during your day.

Photo credit: Thirdman, Unsplash

When you stop social media engagement, even for a few minutes, you stop the addiction process immediately. Even taking a few hours or days without your phone can do wonders for your mood and your self-esteem.

A new study published by the American Psychological Association found that college students who cut their screen time to 60 minutes a day for three weeks immediately began to feel better about themselves and how they look. They reported higher rates of satisfaction with weight and self-esteem compared to those who continue to have unrestricted social media use.

Photo credit: Ayaka Kato, Pexels

Don't worry. You will not be the only one reducing social media use. Others will follow your lead. As more people realize the negative impacts of social media and the positive aspects of reducing its use, you will find others who are aligned with your new, healthier self. You are likely to find increased quality in all aspects of your life with this new way of thinking and doing.

As an experienced social media user, you may have reaped many rewards. And you have also likely learned the pitfalls, the risks, and the downsides of being online. You will see that this knowledge helps to fortify you, giving you what you need for empowerment, breaking free, and letting go.

Stay Complex

Most of us are so much more complex than we may seem on the surface. Let's not cheapen who we are or dumb ourselves down to make other people more comfortable. You are multidimensional and so are your relationships.

Our relationships, all of them, require some degree of active participation. With social media addiction, we tend to take the most important people in our lives for granted. Zen says, "When you love someone, the best thing you can offer that person is your presence." So... put your phone down and pay close attention.

Photo credit: Yan Krukov, Pexels

Yes, personal interaction is more complicated than online interaction. We are just naturally made that way!

Embrace the nuances of speech, body language, and personal histories when you spend time and share physical space with others. Learn to be emotionally intelligent.
Social media addiction can decrease social skills, including empathy and emotional intelligence.

Emotional intelligence is a sign of elevated (high level) thinking and indicates great intelligence, in general. Are you able to perceive, understand, use, and manage emotions? Let's strive to be emotionally intelligent, without hesitation, all the time.

Learn how to empathize, connect, and find common ground with people who are much different than you. This is part of our humanity – the human experience. When you do it, you develop social skills and adaptability. When you observe and use deductive reasoning to understand, you are "observing" instead of just "seeing" and you are using your inherent brain power more fully.

What You Can Do:

- *Go outside and take a walk when tempted by social media.*

- *Limit screen time gradually. Start by reducing by a little each day, and gradually increase the time spent without your phone until you are ready to go without it for at least one day.*

- *Find therapeutic resources, for example, therapists, support groups, counselors, licensed social workers, and psychologists. They can help a lot if you're struggling with addiction of any kind.*

HEALTHY HABITS

"Healthy self" equals "Heal thy self."
Unknown

How do you break old, dysfunctional patterns? By starting new ones that serve you better by helping you stay well. You can learn to develop new coping strategies to increase positive engagement, reduce screen time, and ignore negative, irrelevant, and harmful information. We call these coping strategies "healthy habits." Yes, with motivation and commitment to positive change, you can heal yourself.

It is easier (and tempting) to not attempt to change your life. Going about your day-to-day existence, doing the same thing over and over and over again is very familiar and comfortable. Let's shake things up a bit.

In order to change, a shift in your mindset must occur. See your online usage for what it is. Then, only you can make a decision to do something different. Usually, you must change the energy to take charge of your behavior.

A good first step toward overcoming social media addiction is to develop and practice healthy habits. A digital detox is just the beginning to demonstrate self-control in the midst of chaos: one minute, one hour, one day at a time. While it might be rough at first, keep going. Doing it regularly helps the behavior to become a habit. That doesn't mean that the process will be easy.

Changing your habits is hard work! It means changing what you are used to doing every day. Instead of looking at changes as difficult challenges, you can seize change as an opportunity. It is an opportunity to become self-aware and a better version of yourself. This is how you become even more successful.

A healthy relationship with social media is the goal. That means different things to different people. Generally, it looks like this:

- It doesn't interfere with your positive mental health and self-care priorities.
- It doesn't interfere with your most important offline relationships.
- It is educational and helpful.
- It doesn't impact your emotional well-being in a negative way.
- It doesn't consume too much of your time.
- You can easily jump off and stop your online engagement.

Photo credit: Godisable Jacob, Pexels

Like all of the other choices that you make in your life, you can be smart about how you engage with social media. It takes strength of will, desire, and knowledge.

According to Sarah Coyne, BYU professor of family life, social media can be a positive experience for teenagers and people of any age if they use it well:

"I would love for every BYU student to be mindful about the ways they're using social media, how it's working for their mental health and how it's harming their mental health. Good habits include logging on for a purpose and actively participating rather than passively scrolling, as well as unfollowing those who are exclusionary or have a negative influence."

First, we have to stop denial. The first step to getting better is admitting there is a problem. If you downplay your issue, the time and energy you put into social media, and the pain and hurt that you may feel after social media engagement, you are still refusing to see the problem. This will keep you from getting and staying well.

Secondly, you can learn how to get support. This means:

Removing yourself from isolation, reaching out and being open to receiving concern and care.
Releasing any guilt or shame associated with your issue.
Accepting accountability regarding your behavior.
Being open to changing the way you think about social media and your daily habits.
Being gentle with self-acceptance, understanding that nobody's perfect and living a quality life takes practice, not harsh self-criticism and judgment.
Recognizing that working through stress is healthy and

possible; ignoring or covering it up is not.
- Using tools, for example a journal, to help you process your emotions and get in touch with your feelings.

Thirdly, it means remaining hopeful, knowing that life can be good for you. In a study of over 60,000 respondents from 159 countries, hope, spirituality, zest, gratitude, self-regulation, curiosity, and appreciation of beauty have been shown to increase quality of life, positive health, and purpose.

Photo credit: Armando Castillejos, Unsplash

Wellness is a focus on self, fun, and joyful alternatives.

Healing can take many paths. Recovering from any addiction starts with the acceptance of responsibility and accountability from the person who is in recovery. Your attitude, motivation, and willingness to do better drives this self-directed process.

Hope
Hope is optimistic thinking. Hope helps to control fear. Hope gives you courage to overcome obstacles to healing. Hope can give you more enthusiasm for life. Studies show that the more hope you have, the more positive your health outcomes will be.

Hope often requires a strong foundation of self-esteem; you allow yourself to hope when you know that you are fully capable of handling whatever comes along. Being prepared for the expected and unexpected, being able to adapt. Knowing... you got this! In this way, hope also helps you deal with the anxiety that often comes with the "what-if's" when dealing with something new and different.

Spirituality
There are times in your life when you have to fight *for* your life. In many ways, you are fighting for your spirit. This is not about finding religion. Your spirit is your divine self that knows what's best for you and wants you to know that you deserve so much more.

When your spirit is awake, you can look at media influences knowing that this is not all there is. It's not about anybody or anything else – only you. There's more to you and more life waiting for you to live. Tapping into your spirituality will help you find your highest self.

Peer Support
Meet up with friends! True friends and authentic

connections focus on quality, not quantity. It's nice to see and talk to people in person. Whether your friends are online or off, when your friend group is positive and optimistic, accessible and loyal, they help you stay that way too.

The golden rule is helpful here: be the friend that you hope to have. Try to go to new places to see new things, one-on-one, with people you like. This positive peer support is critical for life and especially for recovery efforts.

Culture
Reunite with family and reignite traditions! Your culture can be a protective factor in helping you adapt to the new normal with technology. It can ground you in a firm foundation of pride and context and bond you with a common understanding of values, what is right and wrong, and appropriate behavior.

When you connect to your culture, you can check in regularly with family, friends, and colleagues on a deeper level than most. Spending time with your peers, surrounding yourself with your culture means affirming yourself with people who love you.

Zest
Start a new hobby! Starting new activities focused on something that makes you happy activates your zest for life. We all have a zest for life. The bliss that we experience where we lose ourselves in being creative can be very personally rewarding.

Zest for life is different from the foggy, dissociative trance that social media often produces. It is more active engagement, conscious movements and actions with a clear, tangible outcome.

When you remove some of the technology around you, you can see what can be gained from being fully present in other areas of your life.

Gratitude

Being grateful is a state of mind. It acknowledges the life we are living in the here and now in a very positive way. Being in the moment and appreciating what we have. That is what mindfulness is all about.

Sometimes, we get so caught up in our response to what someone just said or did that we forget to have perspective, we forget to see the big picture. And when we forget to see the big picture, we focus on little imperfections. So often, we forget to see all of the positive aspects of our entire lives. Maybe you forgot to see how beautiful you are because someone criticized your nose on a post.

Keeping gratitude alive means consistently acknowledging your blessings. You may find that it is helpful to do this with a gratitude journal. A gratitude journal can help you remember the great experiences, even when you are in the middle of something painful.

Self-Regulation

Self-regulation is what you decide to do with yourself. Every day. Self-regulation can help you identify your divine gifts and fulfill your purpose in life. It means consistently acting in your values and your own best interest. Self-regulation means maintaining control of your thoughts, emotions, and daily behavior.

Photo credit: Energepiccom, Pexels

While we strive for positive and healthy self-regulation, there may be times in our lives when we don't quite feel in control, for example, how we can sometimes get stuck in our feelings or triggered by a trauma.

Most of the time, we can work our way out of a rut or any negative thinking with self-compassion, time, and perspective. Other times, we need support, like speaking with friends and members of our family. Still, there are times when you may require more.

Healthy self-regulation also means knowing when to get professional help. Psychiatrists, psychologists, therapists, counselors, and support groups can help you be seen, heard, acknowledged, and understood in ways that social media cannot.

Curiosity

Curiosity is an example of having enthusiasm for life. It

gets you excited about the present and the future. This is especially true about exploring different aspects of your multidimensional self.

Instead of running away, you can turn inward. And expand. Expand your world, expand your mind. Open up to the possibilities of your existence by maintaining curiosity about yourself and the world around you. This means exploring new places and making efforts to have new experiences.

Beauty Appreciation

Beauty is positive energy. Being in a positive environment means being around beauty in your everyday surroundings. This helps you keep a very uplifted mood.

Photo credit: Henri Pham, Unsplash

Your bedroom and your home should be filled with things that make you feel good when you are around them and purify the air: art, furniture, flowers, plants, soft fabrics (for example, cotton and cashmere), pillows, rugs, light colors (for example, pastels like yellow and pink), candles, herbs, and incense for aromatherapy are all very helpful to increase beauty awareness.

Include beauty in your self-expression, for example in your fashion, hair, and makeup, and learn to appreciate it in others too.

Instead of staying indoors on the computer or cell phone all day, I encourage you to go outside to appreciate the infinite beauty of nature, buildings, temples, and parks.

Deinfluencing

There is a backlash to influencers as the presence of deinfluencing has emerged to help neutralize this dominating marketing force. Comprised mostly of Gen Z, deinfluencers are clearly expressing a desire to move beyond social media overconsumption. These social media sensations are telling users what NOT to buy. Deinfluencing activities may provide a healthy balance on specific sites.

Moderation

Moderation is the key. Try these healthy habits:

- Filter your accounts and notifications.
- Try to stay off social media at least one day per week.
- Turn off social media notifications.
- Embrace positivity.
- Limit your social media feeds to include only content that helps you feel good about yourself.
- Ignore social media trolls and energy traps.
- Prioritize travel to distant and not-so-distant

lands to check the vibe and do something different.
- Allow yourself to be in your feelings and embrace them all, good and bad, even if it's scary.
- Be open to learning new techniques for treating social media addiction.
- Limit phone use to 1-2 hours per day. Set your timer or use an app to help you stay on track.

I hope that this Girls Guide is a resource for you so that you know that you are not alone.

What You Can Do:

- *What does it mean to you to fully embrace your awake adult life?*

- *Make a list of 10 things that you can do instead of being on social media.*

- *Think about ways in which you can use your recovery from social media addiction to share with, teach, and help others.*

REFERENCES

Allen, Christie. February 3, 2021. 10-year BYU study shows elevated suicide risk from excess social media time for young teen girls. Available at: https://news.byu.edu/intellect/10-year-byu-study-shows-elevated-suicide-risk-from-excess-social-media-time-for-young-teen-girls

Aubrey, Allison. February 26, 2023. How to help young people limit screen time – and feel better about how they look. *NPR*. Available at: https://www.npr.org/sections/health-shots/2023/02/26/1159099629/teens-social-media-body-image

Barnes, Stephanie. February 11, 2023. 23 examples of self-destructive behavior & how to stop it. *MindBodyGreen*. Available at: https://www.mindbodygreen.com/articles/self-destructive-behavior.

Centers for Disease Control (CDC). February 13, 2023. U.S. teen girls experiencing increased sadness and violence. Available at: https://www.cdc.gov/media/releases/2023/p0213-yrbs.html

Cullen, Katherine. January 28, 2023. How to use social media without losing your mind. *Psychology Today*. Available at: https://www.psychologytoday.com/us/blog/the-truth-about-exercise-addiction/202301/how-to-use-social-media-without-losing-your-mind

Guarnotta, Emily. March 17, 2023. Social media addiction: Could your scrolling habits be a problem? *GoodRx*. Available at: https://www.goodrx.com/well-being/behavioral-addiction/addicted-to-social-media.

Haidt, Jonathan. November 21, 2021. The dangerous experiment on teen girls: The preponderance of the evidence suggests that social media is causing real damage to adolescents. *The Atlantic*. Available at: https://www.theatlantic.com/ideas/archive/2021/11/facebooks-dangerous-experiment-teen-girls/620767/

Johnson, Kristen. September 15, 2021. Instagram taking its toll on teen girls' mental health: Report. *NBC Connecticut*. Available at: https://www.nbcconnecticut.com/news/local/instagram-taking-its-toll-on-teen-girls-mental-health-report/2582713/

Kenton, Will. December 5, 2022. What is groupthink? Definition, characteristics, and causes. *Investopedia*. Available at: https://www.investopedia.com/terms/g/groupthink.asp#:~:text=Groupthink%20is%20a%20phenomenon%20that,of%20a%20group%20of%20peopleb

Madden, Mary, Lenhart, Amanda, Cortesi, Sandra, Gasser, Urs, Duggan, Maeve, Smith, Aaron, & Beaton, Meredith. Pew Research Center. May 21, 2013. Teens, social media and privacy. Available at: https://www.pewresearch.org/internet/2013/05/21/teens-social-media-and-privacy/.

Mayo Clinic. Dissociative disorders. Available at:

https://www.mayoclinic.org/diseases-conditions/dissociative-disorders/symptoms-causes/syc-20355215

McQuate, Sarah. May 22, 2023. 'I don't even remember what I read': People enter a 'dissociative state' when using social media. *University of Washington News.* Available at: https://www.washington.edu/news/2022/05/23/people-enter-a-dissociative-state-when-using-social-media/

Pandey, Erica. February 12, 2023. The rise of the anti-influencers. *Axios*. Available at: https://www.axios.com/2023/02/12/tiktok-instagram-deinfluencing-trend

Perry, Louise. March 3, 2021. Suicide rate among girls higher than ever as toxic social media becomes all-consuming. *New Stateman.* Available at: https://www.newstatesman.com/science-tech/2021/03/suicide-rate-among-girls-higher-ever-toxic-social-media-becomes-all

Staloch, Laura. February 20, 2023. Women who engage in strategic self-presentation online may be more likely to develop characteristics of self-objectification. *PsyPost*. Available at: https://www.psypost.org/2023/02/women-who-engage-in-strategic-self-presentation-online-may-be-more-likely-to-develop-characteristics-of-self-objectification-68287

Weziak-Bialowolska, Dorota, Bialowolski, Piotr, Niemiec, Ryan M. April 2023. Character strengths and health-related quality of life in a large international sample: A cross-sectional analysis. *Journal of Research in Personality*. Available at: https://www.sciencedirect.com/science/article/pii/S0092656622001519.

ABOUT

Girls Guide is a how-to book series for inspiration, self-development, and improvement, in the spirit of sisterhood. The book series is a set of resources created by Angela D. Coleman, Founder & President of Sisterhood Agenda.

EMAIL

sister@sisterhoodagenda.com

IMAGES

Pictures, images, and other photographs are provided by Shutterstock, Unsplash, and/or used under public domain unless otherwise noted.

ACKNOWLEDGMENTS

Thank you to all of our readers, global partners, and to everyone committed to educating, supporting, and empowering women and girls around the globe.

Journal

I love myself today and every day.

Made in United States
Troutdale, OR
11/03/2026

56844711R00046